Copyright Notice

The Dutchmen of Noss

an ornithological expedition to Shetland in 1970

by Jonathan Wills

Copyright © Jonathan W. G. Wills 2023

All rights reserved.

No part of this publication may be reproduced, used or transmitted in any manner whatsoever, including by information storage and retrieval systems, without the express written permission of the author, except for the use of quotations in a review.

Jonathan W. G. Wills has asserted his right under the Copyright, Design and Patents Act 1988, to be identified as the Author of the main text of this Work.

Photographs in this book are by the author, unless otherwise stated. The copyright in images by other photographers and writers, including tables, sketches, pages of notebooks etc., belongs to the persons attributed in the captions.

Cover photograph by Jonathan Wills: The gannetry at the Noup of Noss.

The Dutchmen of Noss

an ornithological expedition to Shetland in 1970

Jonathan Wills

2023

Contents

Acknowledgements v

The Dutchmen of Noss 1

The 1970 census results 9

Birds of Noss National Nature Reserve 13

A kind parting gift 15

Gains and losses since 1970 16

'An epic undertaking' 21

Keeping in touch 23

Postscript: The Bressay Raingeese 24

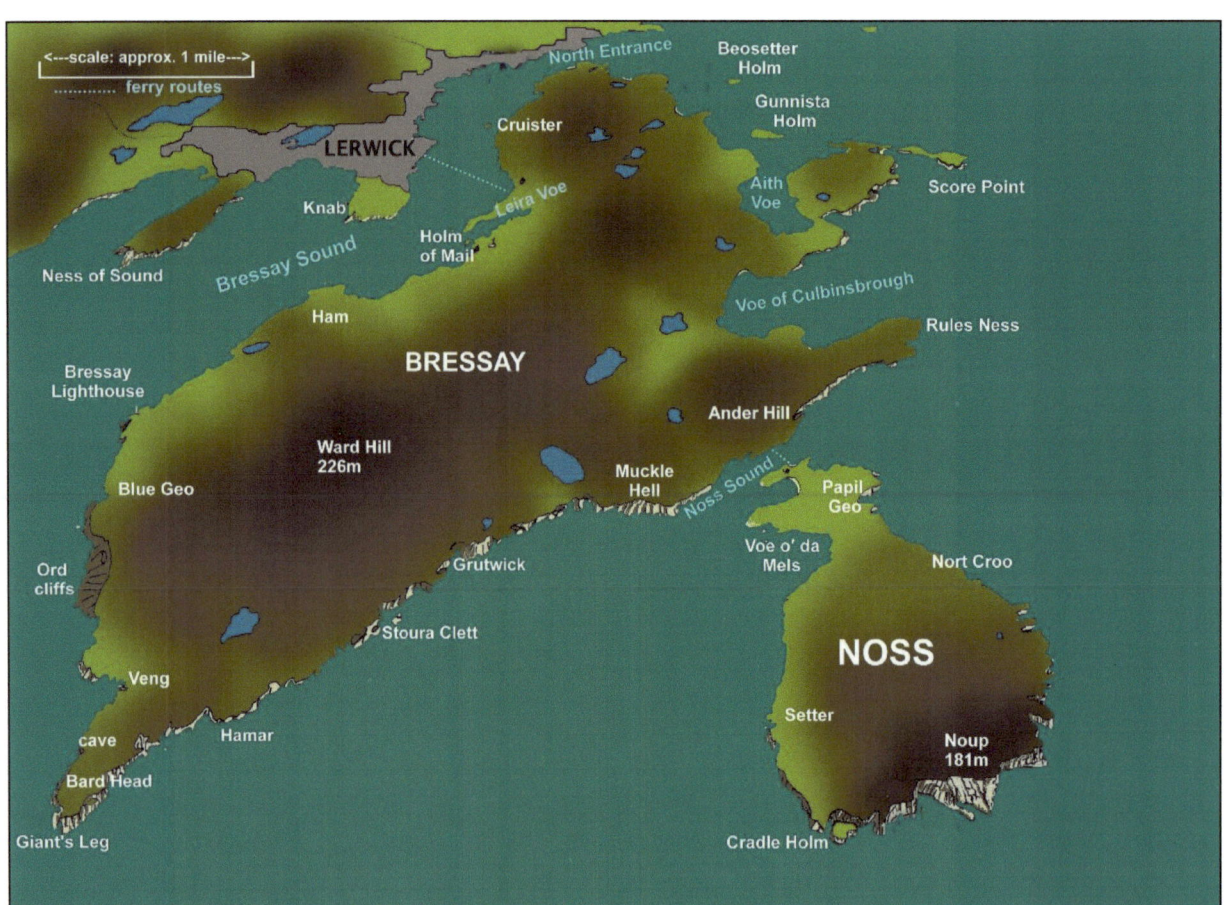

Acknowledgements

One of the pleasures of old age is rummaging through stuff that has accumulated for almost 40 years at the top of our house. There are boxes labelled 'Books – do NOT dump', whose contents I have long forgotten. Other boxes contain notebooks, letters, press cuttings and reports on oil pollution, marine biology and ornithology, plus dozens of faded photographs – all in no particular order. At each spring cleaning certain members of our household describe this collection as junk and threaten to chuck it out. I regard it as my archive, a quarry from which I hew material for my geriatric memoirs. Without it I would have only my faltering memory. So thank you kindly, O Loft of My Life. And thank you also, dear Lesley, for putting up with it, even if you don't accept my argument that all this jumbled paper makes excellent loft insulation.

Thanks are due also to Marcel Groenendaal's loft office in Arnhem. Unlike me, Marcel keeps his papers in good chronological order. I could not have written this memoir without his help. His friendship across the North Sea has been a boon.

The Noss bird census of 1970 was Jan den Held's idea. His memories are bright and his mind as scientifically sharp as ever. It's been a great pleasure to meet Jan again and to resume our correspondence after too long a break.

My fellow former Noss nature reserve warden, Pete Kinnear, has been generous with his time and expertise, particularly in putting Jan and Marcel's data into historical context. Pete's help with entering the data on spreadsheets was invaluable, as was Jonathan Swale's at NatureScot.

Martin Heubeck, who ran the Shetland Oil Terminal Environmental Advisory Group's seabird monitoring programme in Shetland for over 40 years, kindly reviewed my early drafts, saving me much embarrassment. Martin may have missed one of his vocations, for he is a superb proof reader. I am much indebted. Any remaining scientific howlers, typographical errors and arithmetical glitches are my own unaided work.

Tracing Peter Stanley's field notes of the *Operation Seafarer* visit to Noss in 1969 turned out to be a problem but we got there in the end, thanks to Tim Dunn of the JNCC and other helpful folk at NatureScot's Lerwick office and the UK Seabird Group.

Figure 1 The cliffs of Noss seen from the north-east side on a summer's morning. Photo: J. Wills.

The Dutchmen of Noss

In the early morning of Friday 26th June 1970 two Dutch students, Jan den Held and Marcel Groenendaal, arrived in Lerwick, Shetland, after a rough crossing on the overnight ferry *St Clair* from Aberdeen. After buying stores in the town they boarded the Bressay launch *Brenda*, which landed them on the Mail Pier near the Bressay shop and post office. From there they hiked with their rucksacks and tent 2½ miles across the hilly moorland road to Noss Sound, by which time the weather had improved a little. To attract the attention of the Noss boatman, 150 metres away on the other side of the channel, they banged a stick on an old oil drum that hung on a post above the beach. The boatman was me and, as usual, I was having trouble with my Evinrude outboard motor. It was a very busy day with a lot of visitors, including a school party from Lancashire, so I was not in the best of humours when I eventually set aside the outboard motor instruction manual and rowed the Noss boat – a 16-foot, double-ended Shetland skiff – across the tide stream to take the Dutch boys into the island where they would spend the next three weeks.

They produced a letter of introduction from the Nature Conservancy[1]'s Shetland officer, J. Laughton Johnston, in which he explained that they had his permission and approval to study the Skooti Alins and Bonxies (Shetland dialect names for Arctic and Great Skuas) that nested in the Noss National Nature Reserve. I showed them where to pitch their tent close to the farmhouse, near the hand-cranked pump that provided fresh water from the well. That evening, my diary says, I took Jan and Marcel in the boat around the north-east point of Bressay and into the Voe of Culbinsbrough. At the ancient churchyard there they inspected the gravestone of a Dutch East India Company captain, Claes Jansen van Bruyn, who died in 1636 while his ship, the *Manhaetten*, was quarantined at anchor after being refused entrance to Bressay Sound because of fever on board. We got back to Noss Sound in the 'simmer dim' (dialect for midsummer twilight) at midnight. The engine was still causing trouble but despite that we then went off to a fishing boat on Da Sooth Sand to get a 'fry' of haddock. After this we sat talking by the wood-burning stove until two in the morning.

---

[1] Although the Royal Society for the Protection of Birds (RSPB) had paid the tenant of Noss as 'bird watcher' since 1905 and was my employer as warden, the Nature Conservancy was also involved. My employer as Noss boatman was neither of these bodies but the Shetland Tourist Organisation.

Jan and Marcel explained that they wanted to make a whole-island count of the breeding birds, in addition to their work on the skuas. We discussed how it might be done and they decided to walk the whole coast of Noss the following day to work out how to divide the shoreline into manageable *vakken* – sections for counting nests – and to see whether it would be possible to cover all, or at least most, of it from viewpoints on land.

This was an ambitious project. It was quite late in the season for counting some species; for example the Guillemots and Razorbills whose chicks would

Figure 2 Jan and Marcel's tent on the dunes at Noss, June 1970. Photo: Marcel Groenendaal.

already be leaving the cliffs at the beginning of July, while non-breeding immature birds hatched in previous years would be 'rafting' around the coast and exploring the rock ledges on the cliffs. One-year-old Gannets were returning from their winter wanderings off the coast of Africa, to join the 'clubs' of immature birds less than five years old, gathering on rock shelves below the main gannetry, trying their luck at mating displays and prospecting for future nest sites. This was peak season for seabird activity at Noss and the comings and goings of thousands of birds meant that numbers present could vary from day to day and indeed from hour to hour.

We were aware that there had been a survey of Noss the previous year, as part of the national *Operation Seafarer* count of almost all the main British seabird colonies, but in June 1970 we didn't know who'd carried it out and we hadn't seen the 1969 results. *Operation Seafarer* was the first census of the Noss birds since Richard Perry's work in 1946, when he spent the summer making detailed studies of the Skuas and Gannets[2]. Perry also wrote[3] (in prose that was always scientific but also sometimes lyrical) about the then huge colony of Noss Kittiwakes, numbering perhaps 40,000 birds. He described the courtship of the Black Guillemot or Tystie and compiled an account of how the numbers of the various Noss breeding species had varied over the previous two centuries.

However, in his five months on Noss Perry did not try to count the whole island. That was the far more difficult task that Jan and Marcel were now attempting 24 years later – and 15 years after Noss had become a National Nature Reserve. The sheep-farming tenants of the island from 1939 to 1969, the Sutherland family, had, like their predecessors the Jamiesons, acted as bird protection staff for the RSPB and kept some ornithological records but their main role had been to ferry in visitors, forestall egg collectors and prevent human disturbance to rarities such as the Skuas.

When Noss was designated a National Nature Reserve (NNR) in 1955 there appears to have been no prior attempt at a comprehensive survey or census of all the island's birds. In that year the Nature Conservancy did organise a count of Great Skuas (220 pairs), Arctic Skuas (25 pairs) and Great Black-backed Gulls (200 pairs), and in 1958 a survey of Great Skuas only (115 pairs). Then there was an 11-year gap until the *Operation Seafarer* project in 1969 recorded 4,300 pairs of Gannets, 141 pairs of Shags and about a thousand Puffin pairs. That was it. It appears that NNR status, first proposed in 1949, was agreed on the recommendations of various ornithological experts, presumably including Richard Perry, James Fisher, R. A. Richardson, George Waterston of the RSPB and Julian Huxley (who had visited Noss in 1939). All were well-connected 'chaps' who argued – correctly – that the island merited such status, but they did so without any scientific data to substantiate their views. Nor did the British Government ask for any.

---

[2] Perry's methodology was later questioned by the great Gannet expert Bryan Nelson, so the 1946 figures should perhaps be treated with caution.
[3] Perry, R. 1948. *Shetland Sanctuary - Birds on the Isle of Noss*. Faber & Faber, London. This remarkable, highly informative and entertaining book is a classic, despite Nelson's reservations.

So Jan den Held and Marcel Groenendaal were doing important and pioneering work, even though it was entirely on their own initiative and at no cost to the Nature Conservancy Council or the RSPB.

In October 2022, more than half a century after he camped in Noss, Jan recalled how the expedition came about:

"After visiting Ailsa Craig with some friends in 1968, I heard from someone else about Noss and asked Marcel and others if they would like to go there, not only for birdwatching (of course especially the Skuas fascinated us!) but to carry out a census of all breeding [sea]birds on the island."

Figures 3 & 4 Jan den Held in the reed beds at Nieuwkoopse Plassen about 1970 and (right) at home in the Netherlands today.

Jan, who would later become a professional ecologist, already had some experience of scientific ornithology:

"We went to Shetland at a time in our lives when bird censusing in the Netherlands was just beginning to develop. I had done a breeding bird census over several years of one of our wetlands – Nieuwkoopse Plassen, a moor and fen area, with around 1,000 pairs of Sedge Warblers and good populations of Purple Heron, Little Bittern, Savi's Warbler and so on."

Jan's field notes have been lost but Marcel's 1970 notebook still sits on the neatly organised shelves of the office at his home in Arnhem, along with all his other notebooks from a long scientific career. The Dutchmen got to work at once. Within the first couple of days they had divided the coastline into 13 manageable sectors for the census, trying to maximise the area of cliffs they could count from the land.

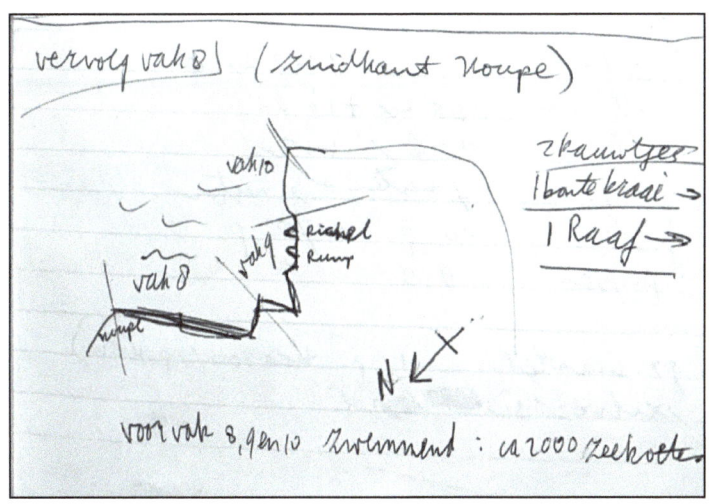

Figure 5 Marcel's notebook sketch of sectors 8, 9 and 10 on the south side of the Noup cliff, with about 2,000 Guillemots on the water and a Hooded Crow, two Jackdaws and a Raven in flight.

Figure 6 Marcel with his notebook in Arnhem, September 2022. Photo: J. Wills.

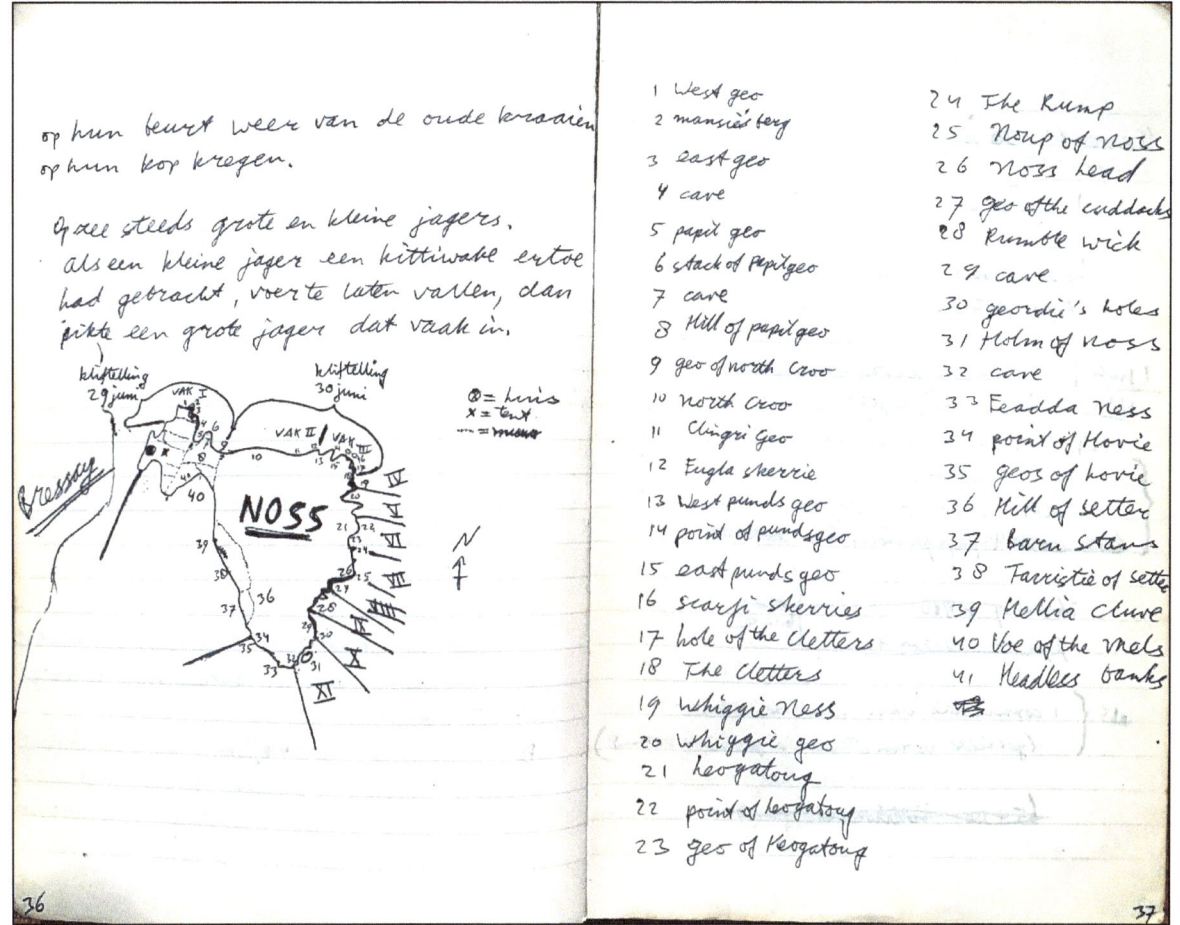

Figure 7 From Marcel Groenendaal's notebook: the initial sketch map and list of place names and survey sectors after the first couple of days on Noss in June 1970. NB: by 30th June the north coast 'kliftelling' [cliff counting] was complete.

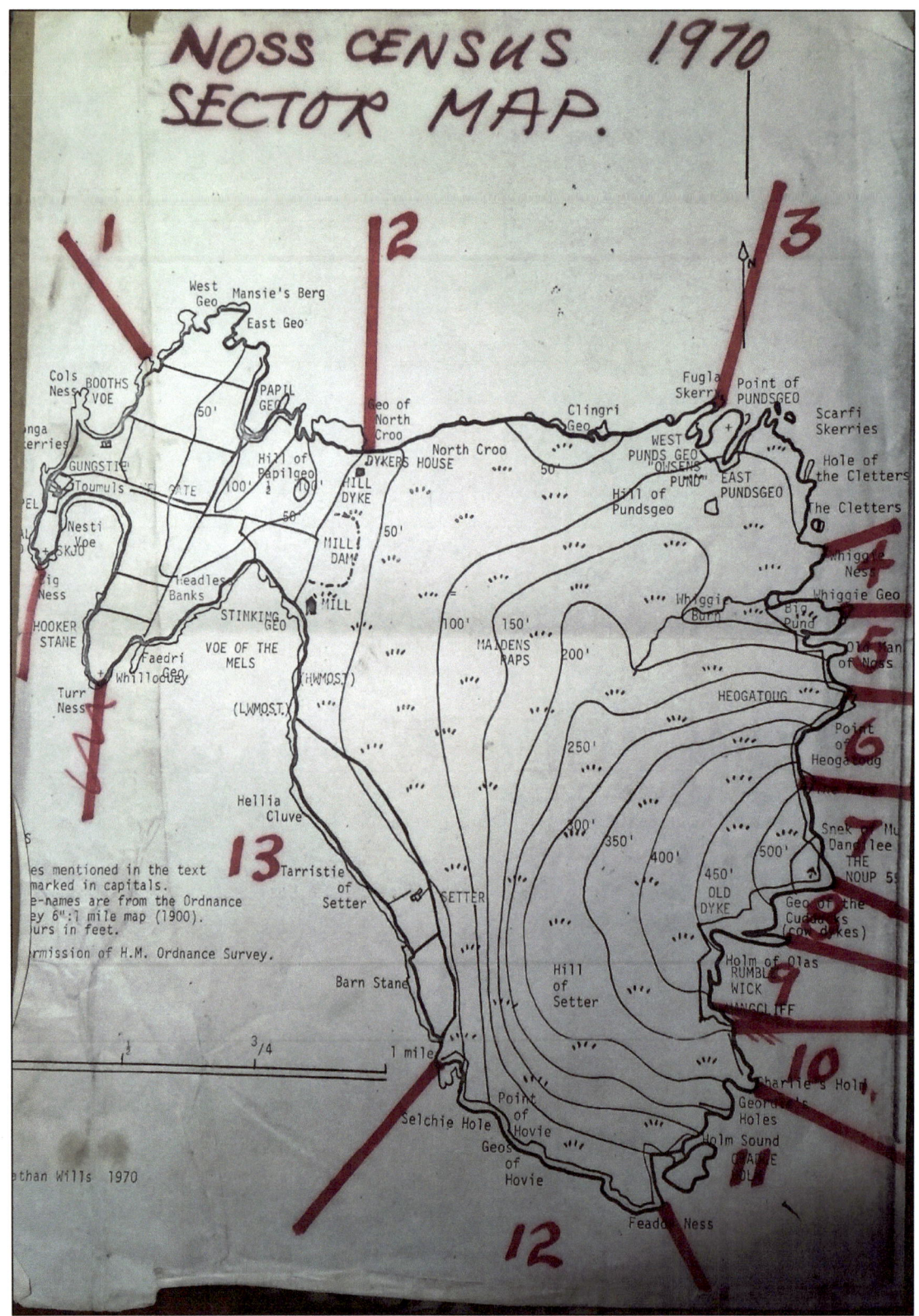

Figure 8 The 'fair copy' of the map showing the sectors for the counts.

Figure 9 Gannet galleries on the south face of the Noup of Noss, with well-grown chicks. Photo: J. Wills, August 2005.

Jan remembers how they went about it: "Counting the breeding seabirds on the cliffs was actually, despite the teeming masses of birds, rather straightforward – if you were patient enough. It was usually easy to divide the cliff faces into smaller segments defined by ledges, rock colour and texture, and then make an accurate count of the birds in each segment. It was much more difficult to make a reliable estimate of the number of Bonxie territories, which were scattered over most of the inner and rather featureless part of the island. We had bought a can of white paint in Lerwick especially for this, so we might divide the area into sections using white painted driftwood posts. However, this proved ineffective and in the end we only managed a rough estimate for this species. The Arctic Skuas were easier because they were concentrated in a small area just east of the 'neck' of the island. One day during field work an attacking Bonxie surprised me, coming from the rear and hitting me a tremendous blow with his outstretched foot on my left ear. I went straight down and was dizzy for the next ten minutes."

Figure 10   Count that! Myriads of seabirds seen from a boat on the east side of the Noup of Noss. Photo: J. Wills, 2015.

On 29th June the evening was calm and clear when I took Jan and Marcel right around Noss by boat, to check on sections that they couldn't see from the land, such as the seaward side of Cradle Holm and the overhanging section of the Noup.

Figure 11 Cradle Holm from seaward, showing ledges hidden from landward viewpoints. Photo: J. Wills, 2005.

The calm weather did not hold. It was temperamental, as so often in Shetland's midsummer. On their fifth night under canvas, torrential rain and a high wind obliged them to abandon the tent and take shelter in the house, sleeping on the stone floor. Forty years earlier the same thing had happened to another Dutchman, the famous Haarlem conservationist and photographer Jan Pieter Strijbos (1891-1983), when he spent three weeks on Noss in 1933[4].

The following day they began work on Sector III at the higher coast beyond the Point of Punds Geo. From his clifftop viewpoint Marcel counted a total of about 700 *zeekoeten*[5] – Guillemots rafting on the sea off the north-east

---

[4] See an account of Strijbos' Noss visits in: Wills, J. *Sixty North to Sixty South – the Lives of Jessie and Tammie Laurenson*. Amazon, 2022. ISBN 9798773034582. pp. 40-58.
[5] 'Sea Coot'.

corner of Noss, along with about the same number of *alken* – Razorbills, and 45 *papegaaiduikers*[6] – Puffins.

My knowledge of Dutch bird names grew rapidly as the young scientists reviewed their notes each evening over supper. Fulmar was the *stormvogel* – the storm bird; the Gannet was *jan van gent*[7]; Tysties were *zwarte zeekoeten* – black sea coots; Kittiwakes *drieteenmeeuw* – three-toed gulls; Herring Gulls *zilvermeeuw* – silver gulls; and Blackbacks were *kleine* and *grote mantelmeeuw*. Nesting along the shore were Shags – *kuifaalscholvers* and Oystercatchers – *scholeksters*, while up on the moorland were Curlew – *wulpen*, Whimbrel – *regenwulpen*, the Skuas *kleine* and *grote jager*, and *eidereend* the Eider Duck.

There were some good 'ticks' among the long days of painstaking work counting thousands of birds of 16 different species: in the evening of 30th June, five days after their arrival, Marcel looked out of the tent to see his first ever *witsnavelduiker* – a White-Billed Diver (*Gavia adamsii*)[8] in the bay of Nesti Voe. Alas, it appeared to be *olie op vleugels en borst* – oiled on the wings and breast – and had red colouring on its legs, perhaps from bleeding caused by ingesting oil as it tried to preen.

The Noss survey was largely complete by 7th July when four friends of Jan and Marcel arrived to join the expedition. It was a fairly rough day when I rowed them into Noss and the next day was worse but the new arrivals were in high spirits. The brothers Michel and Jos Zwarts, and Frans and Sjaak van Rhijn knew how to party as well how to count birds. We were a lively company. On 8th July my diary records:

> *The Dutch boys invited us to eat pannekoeken (pancakes) with them and afterwards we made a bonfire from old wallpaper and all the tarred wood we could find at the west end of the beach [this being unsuitable for the kitchen stove]. Marcel and Michel brought two dozen photocopies of the Noss map from the Library [in Lerwick] so we can now map all their census data. To bed at 2am after coffee and Dutch-English chatter in the kitchen.*

---

[6] 'Parrot diver'
[7] Literally, 'Johnny Gannet' from the Dutch word *gent*, a gander. The old Scots name for the Gannet was the Solan Goose and they are still called 'Solans' in Shetland dialect.
[8] The largest of the *Gavia* family. One or two are seen very occasionally off Bressay. This is the first and apparently only record for Noss.

With the main Noss work finished the results were typed up on my rickety portable and copies sent to the RSPB in Edinburgh and to the Nature Conservancy's Lerwick office, along with our attempts at cartography. On 12th July the Dutchmen set out to survey the eastern and northern coasts of Bressay. Despite frequent rainstorms they managed to visit most of the Bressay lochs over two days and made what was almost certainly the first detailed survey of the island's cliff-nesting birds. All the data were noted down in Marcel's battered and raindrop-stained notebook.

Figure 12 Five of the 1970 expedition members during their visit to Fetlar after the Noss survey was completed. From left to right: Jos Zwarts, Sjaak van Rhijn, Frans van Rhijn (behind Sjaak), Marcel Groenendaal and Jan den Held. Photo: Michel Zwarts.

The 1970 census results

Figure 13 The summary pages of Marcel Groenendaal's 1970 Noss notebook, showing the total for each species of 'klift-vogels' in each of the 13 sectors as pairs or individual birds, plus the Apparently Occupied Nests (AONs) for moorland nesters such as Great Skua, Arctic Skua and Eider. The total was about 62,000 individual birds.

Noss Seabird census July 1970 figures from Marcel's notebook (revised and corrected)																	
sector/vak	1	2	3	4	5	6	7	8	9	10	11	12	13	Totals	Rounded to	Unit	Birds
Fulmar	275	29	355	370	340	535	700	400	760	210	344	360	125	4803	4800	pairs	9600
Shag	10	0	4	11	12	1	4	0	33	1	33	16	0	125	125	nests	250
Gannet	0	0	0	0	0	0	2470	2600	490	2540	80	0	0	8180	8200	birds	8200
Kittiwake	0	0	0	0	440	315	3700	2250	700	1450	895	0	0	9750	9750	nests	19500
Guillemot	0	0	0	1250	6700	6600	1300	5000	2650	1000	2710	280	0	27490	27500	birds	27500
Razorbill	0	0	30	410	250	230	260	150	90	150	121	100	0	1791	1800	birds	1800
Tystie	0	36	14	1	0	0	0	0	0	0	1	0	0	52	52	birds	52
Puffin	120	1	165	130	85	460	320	70	36	30	160	105	0	1682	1700	birds	1700
Total cliff birds																	68602
Herring Gull	90	110	110	0	0	0	0	0	0	0	20	10	75	415	415	pairs	830
GBB Gull	1	1	2	0	0	0	0	0	0	0	200	0	6	210	210	pairs	420
LBB Gull	1	5	5	0	0	0	0	0	0	0	0	0	1	12	12	pairs	24
Common Gull	0	15	0	0	0	0	0	0	0	0	0	0	0	15	30	pairs	30
Arctic Skua															35	pairs	70
Great Skua															160	pairs	320
Total gulls & skuas																	1694
TOTAL ALL SEABIRDS																	70296

Figure 14 Noss sector totals, tabulated by Pete Kinnear from Marcel's notebook.

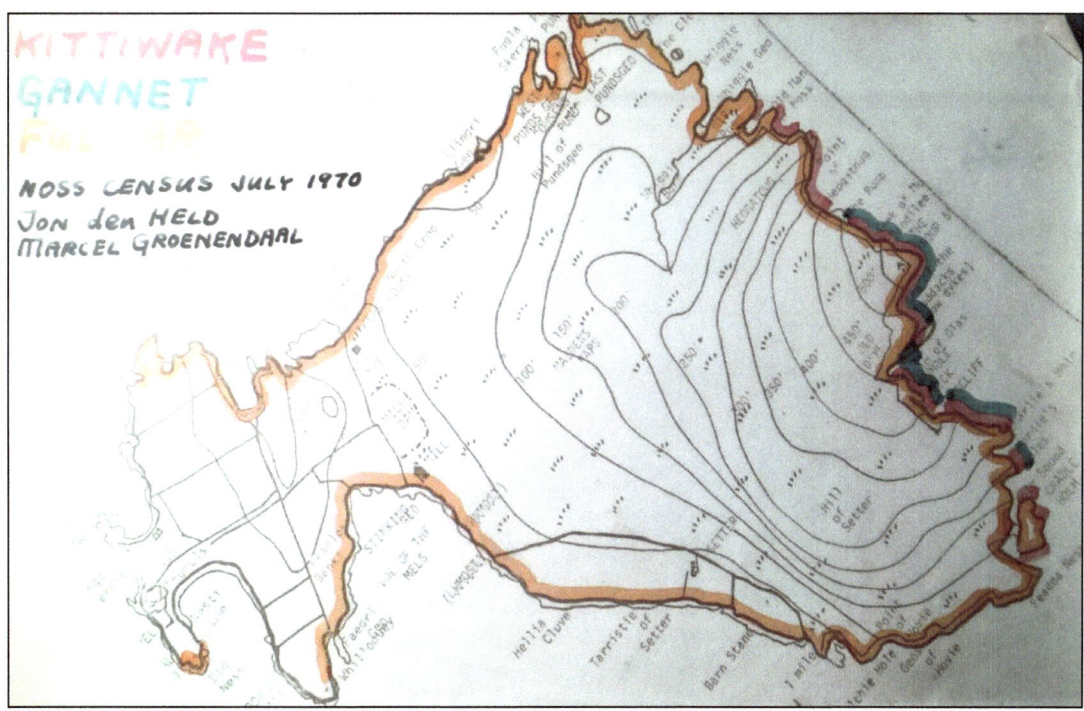

Figure 15 Map showing the territories of three coast- and cliff-nesting species, by Jan, Marcel and the author. In 1970 there was no electronic imaging nor standard A4 paper size so the map had to be angled to fit Noss onto the sheets of foolscap paper that we had available.

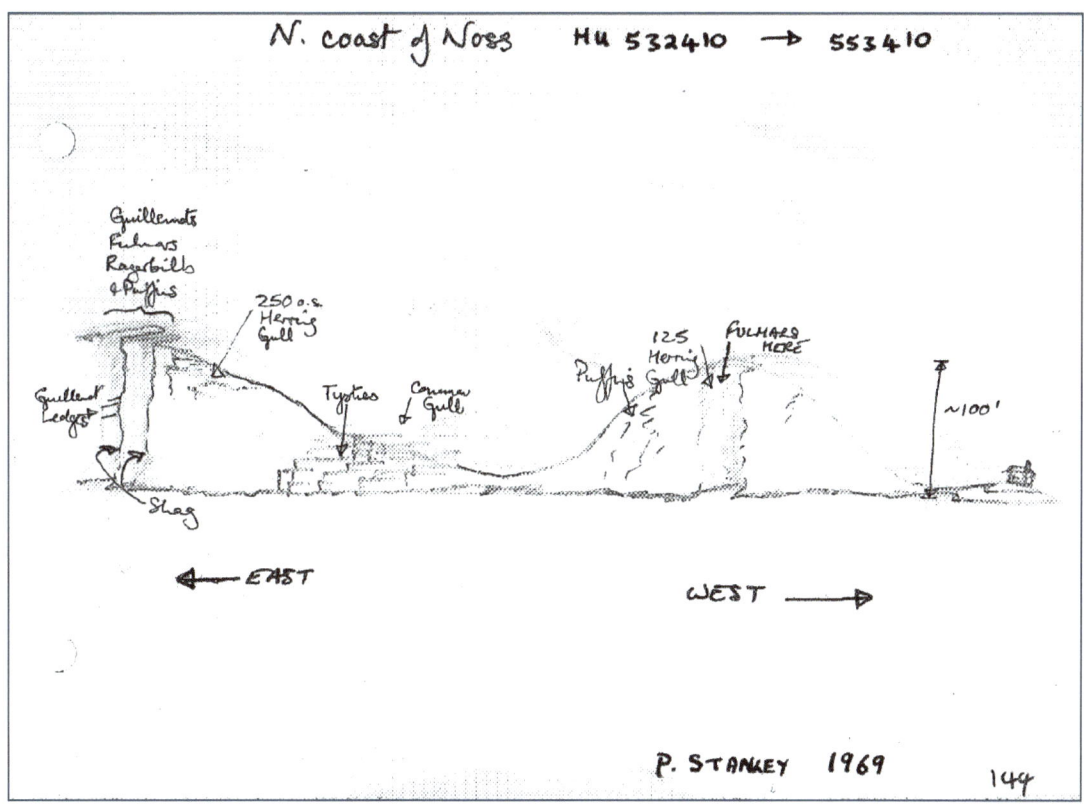

Figure 16 One of Peter Stanley's 'Operation Seafarer' record cards from July 1969, showing the north coast of Noss. Compare this with the more detailed maps and sector counts of the 1970 census in Figures 8, 14, and 15. Reproduced by kind permission of the UK Seabird Group.

Figure 17 A vantage point to count Gannets and Guillemots on the south face of the Noup of Noss. Photo: J. Wills, 2022.

Figure 18 Guillemot ledges, Noss. Photo: J. Wills, 2019.

Figure 19 A 'club' of non-breeding Gannets on Cradle Holm, Noss. Photo: J. Wills, 2022.

Birds of Noss National Nature Reserve

Figures 20 - 26. Top, from left: Noss Fulmars, Gannets and Kittiwakes. Centre: a raft of Guillemots under the Noup. Below, from left: Noss Bonxie, Shag (immature) and Puffins. Photos: J. Wills.

Figure 27 Sjaak van Rhijn attending to his pedicure outside the warden's house, July 1970. Other photos of the expedition members on Noss have been lost, alas. Photo: J. Wills.

Figure 28 The author hauling up the Noss boat before a summer gale, July 1970. Photo: Georges Larondelle.

Figure 29 There was no lack of driftwood on Noss in 1970 for the kitchen stove, to dry out and warm up soaked and shivering Dutch bird counters after a day on the cliffs. Photo: J. Wills.

A kind parting gift

On 17th July 1970 my diary recorded:

> *Up at 6.30am to ferry the Dutchmen across - they are going to Fetlar and then to try and get some work. Three of them may be back. They kindly left me a bottle of whisky.*

Marcel Groenendaal's notebook records their visit to Fetlar, where they saw that island's celebrated rarity – *grauwe franjepoot*, the little Red-necked Phalarope, a globe-wandering species where the male is the dowdy one while the gaudy female leaves him to do most of the incubation. Other Fetlar excitements included hearing young *stormvogeltjes* – Storm Petrels calling from their nest holes. More amazing still for the Dutch birders was watching a flock of 70 *Noordse pijlstormvogel*, Manx Shearwaters, wheeling over the Wick of Tresta, something we are most unlikely to see today.

Jan also remembered: "I think it was on Fetlar that a Great Skua staged a drama for us: the bird caught an Oystercatcher in mid-flight, sat down on it in the water, drowning it in a few minutes, and then calmly started to eat it!"

After some more birding, in Yell, Marcel and the Zwarts brothers returned to the Netherlands but on 23rd July Frans and Sjaak van Rhijn came back to Noss with Jan to camp for a few days, and to lend a hand with my chores as boatman:

> *Jan, Sjaak and Frans helped re-paint the boat, which has got very bashed, with one split plank on the port side... No wonder. A bumpy day on the Bressay shore due to the swell and landlubber tourists...In the evening all four of us decorated the east bedroom with its first coat, discovering in the process a box bed that had been blocked off and papered over for some reason. To bed at 12.30am, very merry.*

Late on Sunday evening, the 26th of July, my Dutch guests left Noss for the last time. I took Jan, Frans and Sjaak in the Noss boat to Lerwick, around the southern tip of Bressay in fading light:

> *A very hairy trip until we reached the lee of the Bard. Nearly swamped by [a fishing boat called] the 'Nil Desperandum', appropriately enough. Engine overheated and stopped twice.*

So much for health and safety at work on the sea. No wonder that after the NCC took over the wardening of Noss from the RSPB a few years later they provided an inflatable boat for the Noss Sound crossing, with two engines, neither of them the evil Evinrude two-stroke outboard which we had somehow survived during that eventful summer of 1970. They also provided lifejackets for all hands, distress flares and a radio[9], these being novelties during my single season as the Noss boatman.

Marcel Groenendaal came back to Shetland in 2005 for the 50th anniversary of the Noss National Nature Reserve. I took him around the island (in some comfort and much greater safety than in 1970) aboard my twin-engined, 12-metre launch *Dunter III*, at that time running the Seabirds-and-Seals wildlife cruises taking visitors to see the huge colony of Gannets, which was more than twice as big as it had been 35 years earlier. We kept in touch and some years later I visited Marcel in the Dutch bird island of Texel, where he worked for many years as one of the wardens. In return for my Noss Gannets he showed me his Texel *lepelaars* – Spoonbills (*Platalea leucorodia*).

In 2019, by then retired, Marcel spent a holiday with us in Bressay. There was a walk around Noss, of course, and much reminiscing about the 1970 census, the significance of which I was at last beginning to understand. I dug out the faded census maps and statistical summaries from a tea chest in our attic and Marcel said he thought he still had his notes. When I visited him at his new home in Arnhem, in September 2022, he showed me some more *lepelaar* (in a reserve near Nijmegen), produced his field notebook for June and July 52 years earlier, and put me back in touch with Jan den Held.

Gains and losses since 1970

Since 1970 there have been big changes at Noss: some birds have greatly increased, such as the Gannets and Fulmars (even after the mortality from the 2022 bird flu epidemic); but more species have declined in numbers, notably Arctic Skuas and Kittiwakes (about 20,000 birds then, only a couple of hundred now); Lesser Black-backed Gulls have gone; the Eider flocks have almost disappeared and the auk family are also fewer in number than 53 years ago; the once dominant Noss Bonxie, whose behaviour has so

---

[9] Laughton Johnston, the Shetland NCC officer, did try out a portable radio in Noss but although we could hear the Lerwick Coastguards on it the set was so low-powered that they couldn't hear us, so a flagpole remained my only means of signalling to the outside world.

fascinated generations of ornithologists, suffered an astonishing 80% mortality in the 2022 bird flu epidemic, whereas the gannets only decreased by 16%. Better news is that Arctic Terns now breed regularly on the island. In 1970 there were none.

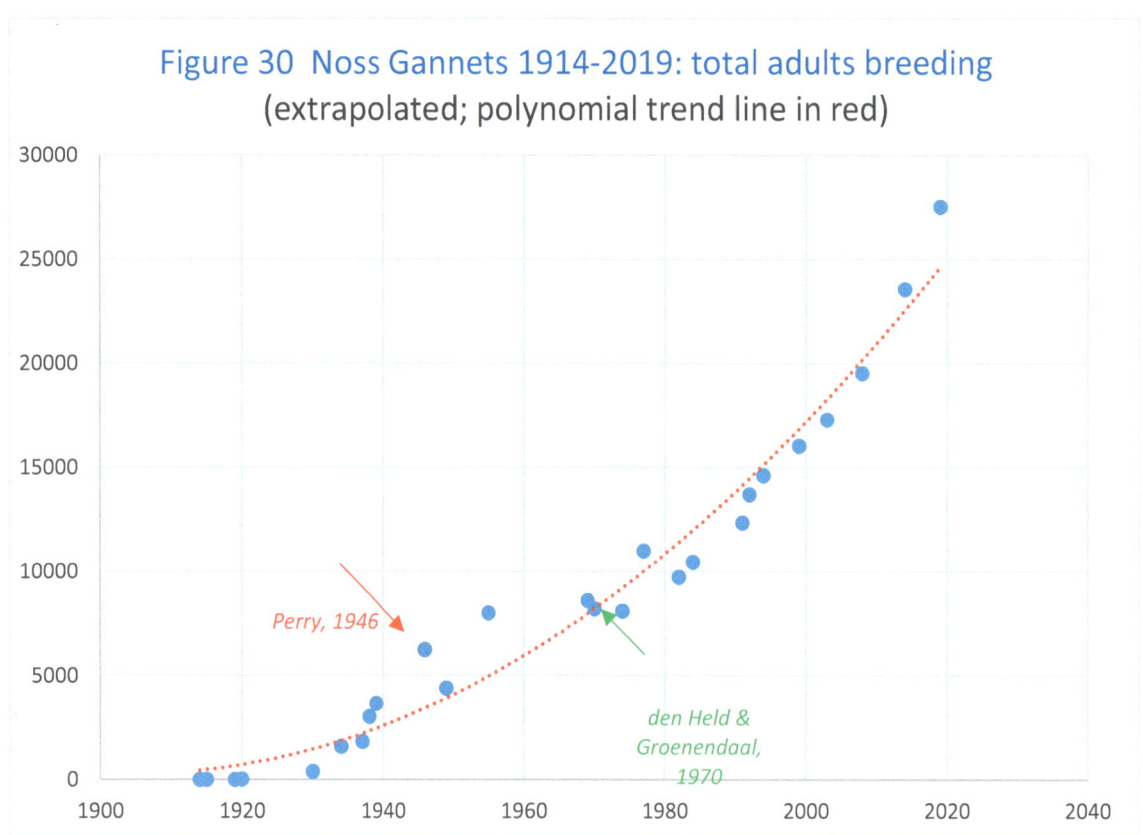

The definitive history of bird recording in Noss was published by the Scottish Ornithologists' Club in 1986[10] as a paper in the periodical *Scottish Birds*. 'Breeding Birds of Noss' included the totals reported to the RSPB and the Nature Conservancy by the Dutch students but, surprisingly, did not attribute them to den Held and Groenendaal, nor do their names appear in the acknowledgements. Perhaps it was a clerical error. Perhaps it was thought that amateurs 'didn't count'. But of course they did count, in both senses.

---

[10] Willcox, Richardson & Dore, 1986. Breeding birds of Noss. *Scottish. Birds* vol. 14, pp. 25-32.
https://www.the-soc.org.uk/files/docs/about-us/publications/scottish-birds/sb-vol14-no01.pdf

We now have a much better idea of what is happening to the Noss bird populations than we used to. This is because for over 40 years governments have allocated resources to manage the island in a manner appropriate to its status as a National Nature Reserve. The Nature Conservancy became the Nature Conservancy Council which morphed into Scottish Natural Heritage which in turn has now been re-badged as NatureScot. More important than the name changes has been a budget to pay the two summer wardens ('seasonal site managers' as they are called in NatureScotspeak). Since 1975 there has been a regular seabird monitoring programme with detailed annual reports and whole-island counts for the main species at five-yearly intervals.

Thanks to the wardens and their colleagues in the Lerwick office, we have a lengthy, reliable run of data, not only for birds but also for the mammals, insects, plants and fungi that make Noss their home. During the Covid-19 pandemic Noss was closed to visitors but monitoring work continued, which is why we have a good idea of the scale of the bird flu casualties.

Figure 31 A Bonxie and a Gannet, both killed by bird flu on Noss. May 2022. Photo: J. Wills.

Bird censuses are notoriously difficult, particularly when observing large, crowded colonies with several species sharing a cliff, as at Noss. Accurate comparisons between records made there at intervals over a period of 150 years are problematic because observers may have been counting different things – for example, individual birds, mated pairs, apparently occupied nests (AON) and, in the case of Gannets, apparently occupied sites (AOS). And of course the counting methods they employ may vary. So the 1970 Noss data, although roughly in line with the long-term trend line, are not fully compatible with modern practice. This is because seabird census techniques were still evolving well into the 1980s, although since 1975 the counts on Noss have followed a consistent, regular methodology, making them more credible and comparable. It is easy to criticise earlier studies but we should bear in mind that modern researchers have high-resolution aerial photographs, smartphone cameras, computers and other electronic aids which enable them to do more work, more accurately and far more quickly than was possible with Jan and Marcel's binoculars and notebooks half a century ago. Earlier estimates were sporadic and sometimes unreliable. As noted above, the figures produced by Richard Perry have been questioned: his Gannet count on Noss in 1946 does look like an over-estimate, out of line with the trend (see Fig. 30 above).

The *Operation Seafarer* count of Noss seabirds, mentioned earlier, was carried out in the first week of July 1969 by a well-qualified scientist, Dr Peter Stanley, on behalf of the Nature Conservancy. He later became Director of the Central Science Laboratory of the UK Ministry of Agriculture, Fisheries and Food. Stanley's methodology was similar to the Dutchmen's and his figures were fairly close to theirs, as the table below shows:

Numbers of seabirds by species recorded on Noss 1969 & 1970					
	1969	1970		1969	1970
BREEDING PAIRS / AONs			INDIVIDUALS		
Eider		41	Eider		
Fulmar	2,080	4,839	Fulmar		
Gannet	4,300	4,100	Gannet		8,200
Shag	141	125	Shag		
Guillemot			Guillemot	24,155	27,500
Razorbill			Razorbill	3,120	1,860
Puffin			Puffin	800	1,700
Tystie			Tystie	52	52
Great skua	210	150-170	Great skua		
Arctic skua	40	30-40	Arctic skua		
Kittiwake	10,510	9,750	Kittiwake		
LBB Gull	30	12	LBB Gull		
GBB Gull	304	210	GBB Gull		
Herring gull	500	415	Herring gull		
Common gull	20	15	Common gull		
Oystercatcher		3	Oystercatcher		
Source	Seafarer - Peter Stanley	den Held & Groenendaal		Seafarer - Peter Stanley	den Held & Groenendaal

Operation Seafarer was a great national endeavour and it produced a remarkable book[11] that was a baseline for later estimates of Britain's seabird population. Surprisingly, although the book used some of Jan and Marcel's figures, it did not include them all and, like the *Scottish Birds* paper, did not attribute the source.

The discrepancies between the 1969 and 1970 census results for Fulmars, Razorbills and Puffins illustrate the difficulties of counting seabirds. Working on his own, Dr Stanley seems to have had only had two days for his counts at Noss. On one of them he was in a hired boat during a south-westerly gale, according to the record cards he completed, copies of which are preserved in the archives of the Joint Nature Conservation Council (JNCC). Although he was surveying the sheltered eastern side of the island that day (3rd July), conditions would have been poor. On the other day (7th July) it was raining and overcast, which may have affected visibility, although he was observing the colonies from viewpoints on land on that occasion.

The 1969 figure for Fulmars was 2,080 apparently occupied nests (AONs) whereas the 1970 total for 'pairs' was more than double that, at almost exactly the same time of year. However, Fulmars spend eight years acquiring nest sites and mates before they breed. In very windy weather the numbers may fluctuate markedly as non-breeding birds desert the cliffs. Because of the high number of non-breeders, counts may also differ significantly at different times of day. So the apparent doubling of the Noss Fulmar population in a single year is probably not what was actually happening.

The figures for Razorbills are almost the reverse, with an apparent decrease of about a third between July 1969 and the same month the following year. Razorbills are always the earliest seabirds to take their chicks to sea and the timing of departure varies according to food supply. It may be that 1970 was a better year than 1969 for the sand eels (*Ammodytes spp.*) that the birds need to feed their chicks; so the birds may have left Noss earlier than in the previous July. Also, rough weather can delay departure and the gale of 3rd July 1969 might have kept chicks on the ledges even if they were ready to go to sea.

The doubling in the number of Puffins counted in 1970 compared with 1969 is statistically insignificant because Puffins, being burrow nesters, are notoriously difficult to count accurately. As with Guillemots, there are usually large numbers of non-breeding birds in attendance. Bad weather may

---

[11] Cramp, S., Bourne, W. R. P, and Saunders. D., 1974. *The Seabirds of Great Britain and Ireland*. Collins, London. ISBN: 0 00 216753 0.

affect numbers on shore and also the availability of food for chicks. The fact that Jan and Marcel had more pairs of eyes available, more time and could therefore afford to wait for better weather will also have influenced the totals.

Even with uncertainties like these we can still make general statements that some Noss bird species have greatly increased and others massively declined since 1970. What is certain is that without the efforts of those enthusiastic young Dutch volunteers as 'citizen scientists' in that summer so long ago we would know much less about the birds of Noss. In the tables and graphs charting changes in the island's wildlife over the past century and a half[12] there is a blip labelled '1969-1970' when, after a quarter century of patchy information following Perry's efforts, all of a sudden we have two major surveys in consecutive years (and on almost the same dates) by observers who understood scientific bird census methods – Peter Stanley's work and the more detailed survey by 'the Dutchmen of Noss'.

Unknown to me, Jan had returned to Shetland in 1974 and counted the Gannets of Noss again. This time he found 8,093 individual birds, slightly fewer than the 8,200 recorded in 1970. Then he went to Unst and camped at Burrafirth, just half a mile from where I was living at the time – but he didn't know that and we didn't meet.

'An epic undertaking'

Pete Kinnear is a professional ornithologist who was warden of Noss in 1973, became a founder of the Shetland Bird Club that year and worked on major Shetland bird surveys before the Sullom Voe oil and gas terminal opened in 1978. He rates the Dutch students' work highly:

"I do regard what they did as an epic undertaking," he writes. "I would never have contemplated trying to count Noss. They were part of a movement throughout Europe to embrace a different way of life which included sustainability, placing nature conservation in the forefront of life, taking on poorly paid roles to further their knowledge and play their part in raising awareness of the importance of natural habitats and wildlife."

Pete agrees that the 1970 Noss census was more detailed than *Operation Seafarer* because the Dutch effort had more observers and more time. It

---

[12] For summaries of the Shetland data see the classic work: Pennington, M. *et. al.*, 2004. *The Birds of Shetland*. Helm County Avifauna. ISBN: 0-7136-6038-4.

would become the baseline survey for all subsequent work on the island. It was the start of something much bigger:

"Oil was the trigger for so much of the research undertaken in Shetland," Pete says. "After *Operation Seafarer*, Mike Harris was despatched north in 1974 to look at what needed to be done to get a better handle on Shetland seabird populations. Bob Furness and the Brathay Exploration Group were doing Foula, University of East Anglia students were on Hermaness in Unst and Jim Fowler began intensive field work in Shetland. People like me took on short term contracts to survey skuas, seaduck and begin the monitoring of Shetland seabirds using the techniques that Mike Harris developed on St. Kilda and the Isle of May. A growing army of well-trained former Fair Isle Bird Observatory staff emerged. Many ultimately gained PhDs... and went on to have long term careers in science, teaching and conservation. They joined the well-informed local population and fought many important battles together: Mike Richardson, Martin Heubeck *('I've only been here 40 years, what do I know?')*, Pete Ellis, Nick Dymond, Iain Robertson etc. all played their part. The formation [in 1976] of the Shetland Oil Terminal Environmental Advisory Group (SOTEAG) played a key role in what happened and what was studied.

"Shetland must be one of the most intensively studied places in the world. We were all post war babies, all of us starstruck by 'remote' places. The twitchers who stayed the course became world-travelled experts who wrote and drew the new generation of field guides and led expeditions to even remoter destinations... and 1970 was a key moment when much of this began, so it is fitting that you should be writing the story of Jan and Marcel and their colleagues, 53 years on."

Reflecting on their ambitious enterprise, half a century later, Jan says: "It is very nice to know that we have done something that is also appreciated by others. And of course there is the amazement that in youthful overconfidence we just started something and also brought it to a reasonably good end."

The data collected by Jan, Marcel and their friends were never peer reviewed or published in a scientific journal. So this memoir has been an attempt to salute their achievement and public-spirited endeavour. In later life they would earn salaries and pensions for their work. On Noss in 1970 they did it unpaid, for love of the natural world, on a diet of coffee, fish and *pannekoeken* (with just the occasional nip of whisky for medicinal purposes).

Bedankt en goed gedaan! Bravo!

Keeping in touch

Figure 33 *Marcel and the author aboard m.v. Dunter III off the Noup of Noss, 2005.*

Figure 32 *Three old-timers: Jan den Held (left), the author and Marcel Groenendaal meet for a reunion lunch and selfie at Aalsmeerderdijk in the Netherlands, 3rd April 2023.*

Figure 34 *Noss from the south-east. Photo: Hellio van Ingen.*

Postscript: The Bressay Raingeese

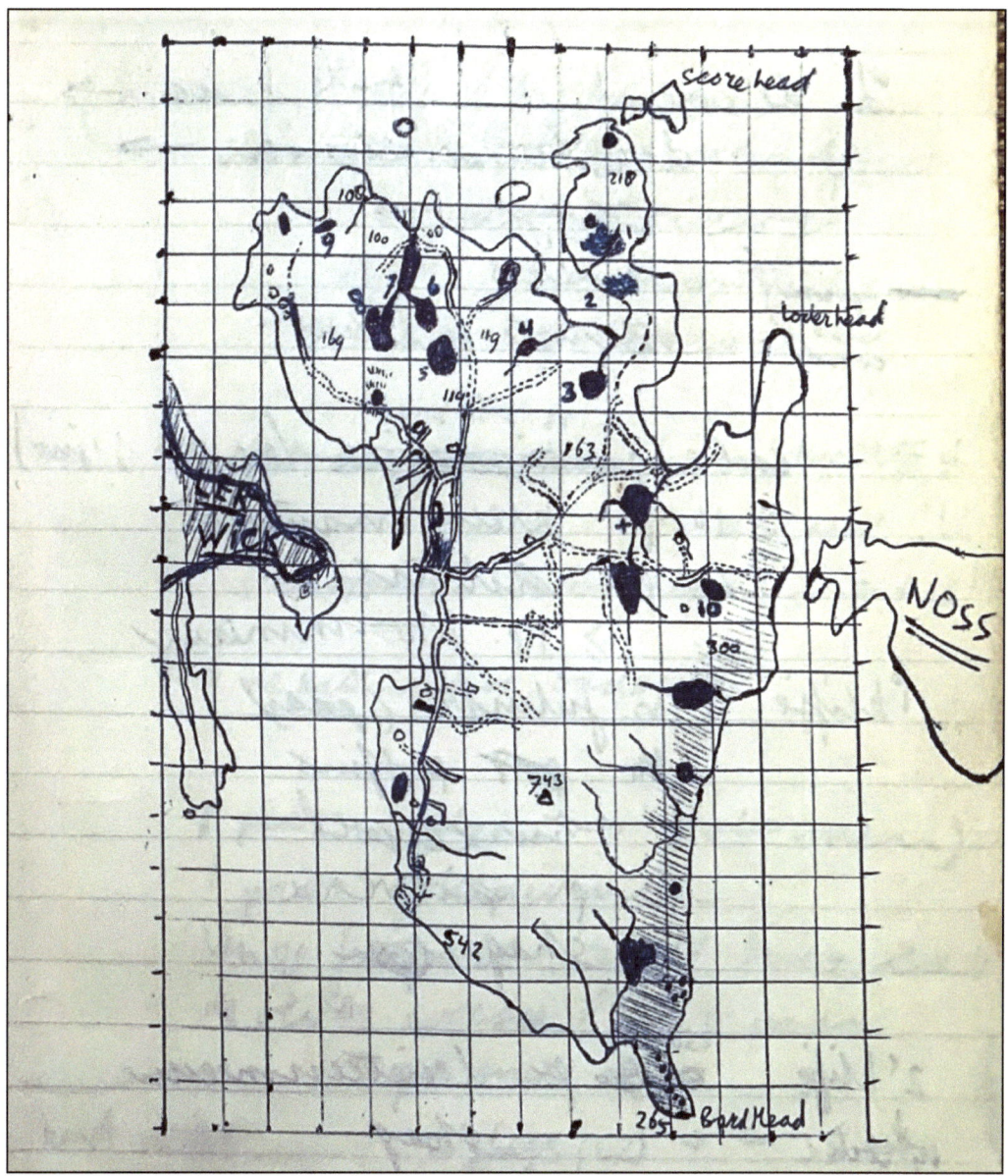

Figure 35 Marcel's sketch map of Bressay, with the lochs numbered.

The expedition's records of the *roodkeelduiker* or Red-throated Diver (*Gavia stellata*, Shetland dialect's 'Raingoose') are of particular interest as there does not seem to be an earlier census of all the lochs where the species is known to have bred in Bressay.

In the second week of July 1970 it was perhaps a little early in the season to census all the Red-throated Diver chicks but the total of a dozen pairs – 27 adults and five juveniles – compares starkly with today's figures.

July 1970: The Dutch students' survey of some Bressay lochs and the eastern coastline of the island (Huntersfield to Bard).
Source: Marcel Groenendaal's field notebook.

Date	Note-book page	No. on MG sketch map	Loch name (on 1880 Ordnance Survey 6":1 mile Map)	Red-throated diver sightings (*Gavia stellata*)
19700712	63	*Not numbered*	Loch of Grimsetter ('large loch in a field')	1 + 2 juv
19700712	64	*Not numbered*	Loch of Seligeo	1 pair + 1 juv
19700712	64	*Not numbered*	Probably Sand Vatn	1 + 1 juv
19700716	67	*Not numbered*	'zuidpunt baai' (probably Hope Wick)	1
19700716	69	1	Deadwall Loch (Aithsness)	1 + 1 juv
19700716	69	2	Aith Voe	8 (individuals)
19700716	71	*Not numbered*	off Beosetter and Gunnista Holms	2 (individuals)
19700716	71	*Not numbered*	Voe of Culbinsbrough	2 (individuals)
19700716	72	3	Loch of Bruntland	3 adults
19700716	73	4	Loch of Aith	-
19700716	73	5	Beosetter Loch South	-
19700716	73	6	Beosetter Loch West	3 adults
19700716	73	7	Beosetter Loch North	-
19700716	73	8	Beosetter Loch West	1 adult
19700716	73	9	Loch of Huntersfield	-
19700716	73	10	Ullins Water	1 pair
	77		**TOTAL (breeding pairs)**	**12 or 13** (total of 27 adults and 5 juveniles recorded)

Over the past half-century the poor Raingoose has suffered severely from greater predation of chicks by gulls, skuas and otters, increased disturbance at the nest from pet dogs and their thoughtless owners, and also from a shortage of sand eels – the tiny fish that is an essential summertime food for Raingeese as well as for Kittiwakes and the three auk species when raising chicks.

In the summer of 1970 there were vast shoals of sand eels around Bressay and Noss, particularly in Noss Sound. Later they vanished. They are slowly recovering, at last, after being severely over-fished in the 1980s, ironically to feed the fish meal and oil factory in Bressay. Despite the belated ban on that disastrous industrial fishery there are now less than half a dozen pairs of Raingeese breeding in the island, at most, and very few of their chicks make it to fledging.

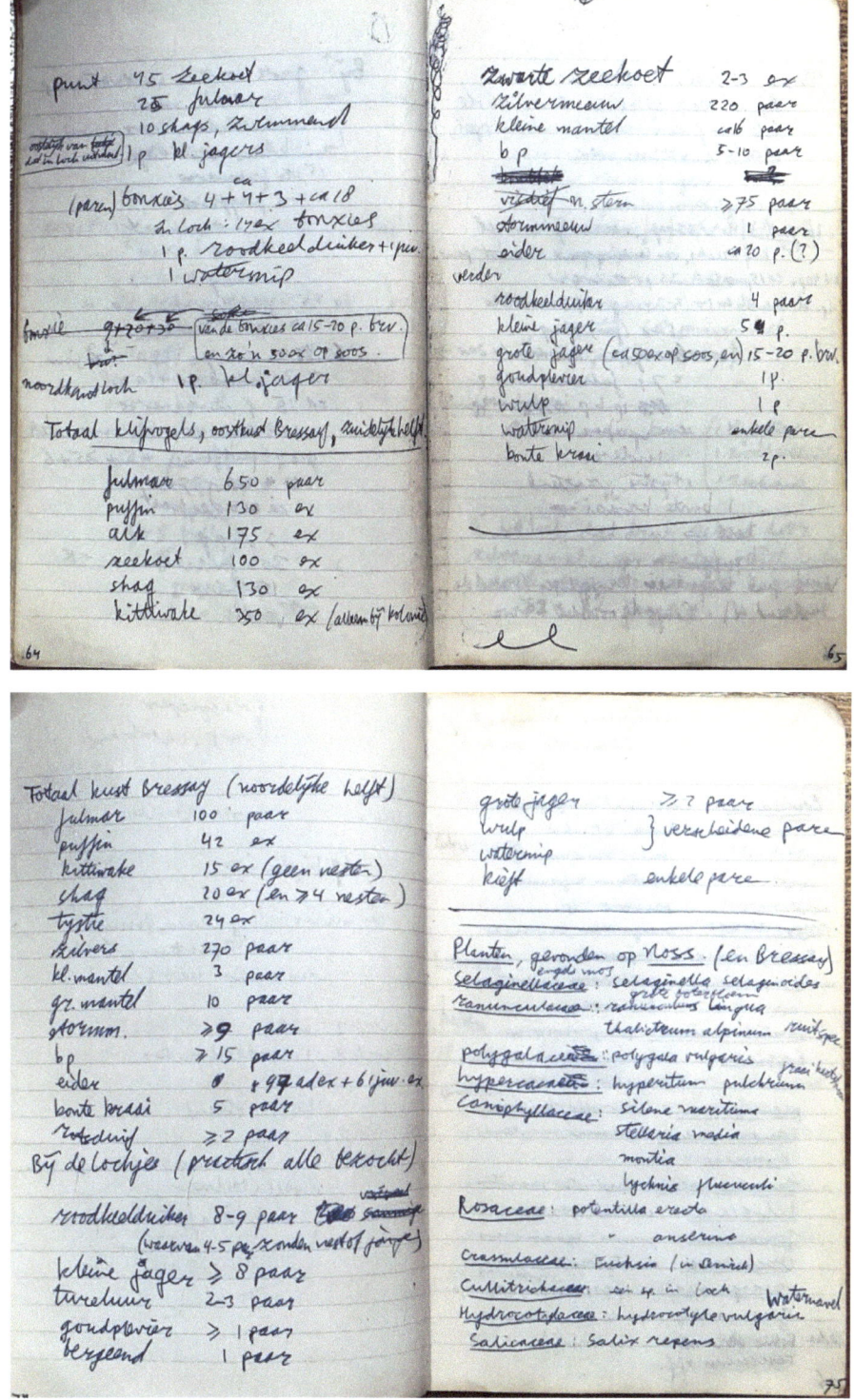

Figures 36 & 37 Total numbers of each bird species for the southern and northern sections of Bressay, with some botanical notes on both islands, from Marcel's notebook.

Figure 39 Red-throated Diver. Photo: David Karnå / Wikimedia Commons

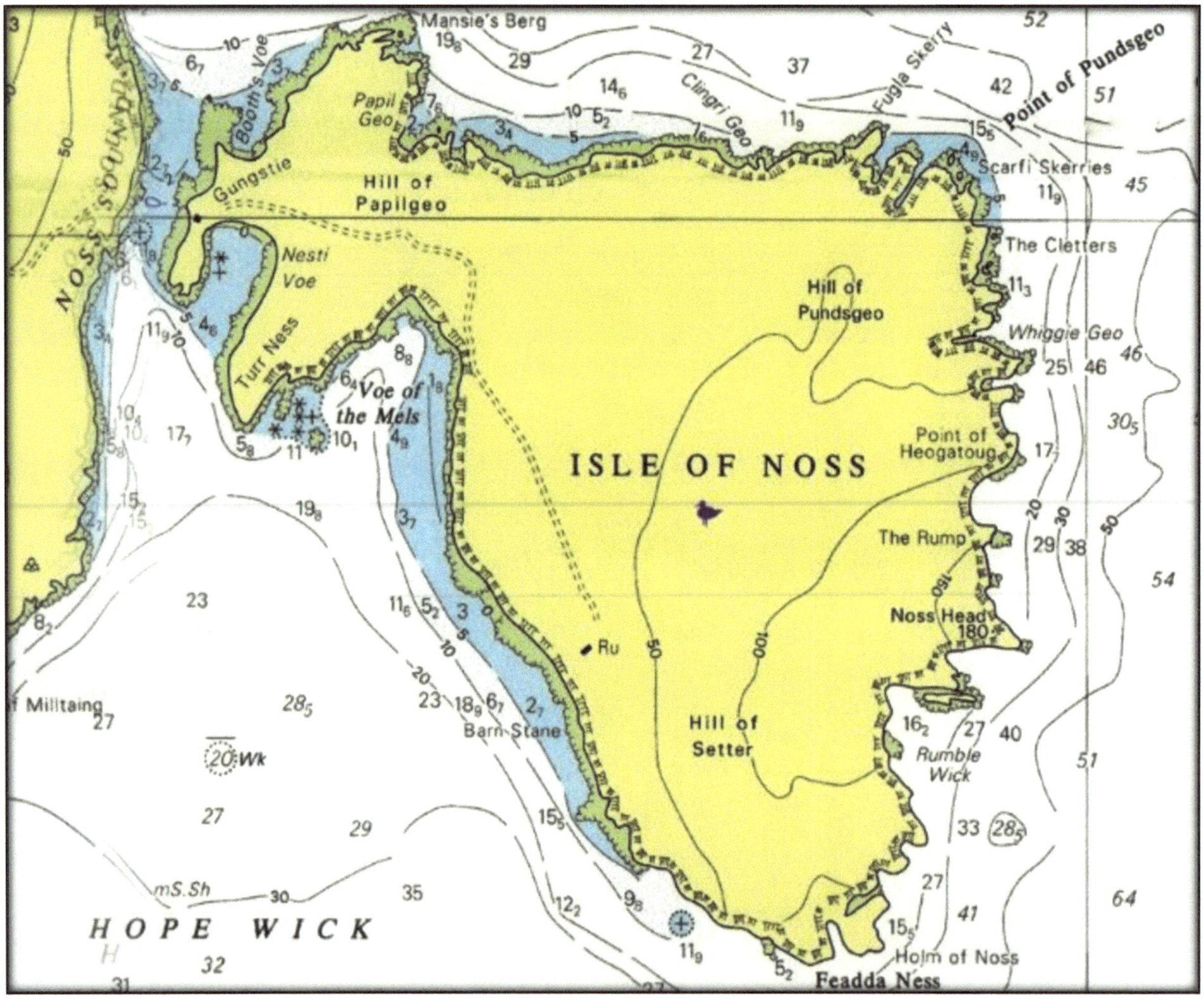

Figure 38 Noss. From Admiralty Chart No. 3272.

Other books by Jonathan Wills

Environment and natural history:

Seabirds and Seals – the stories behind 25 years of wildlife cruises around the Shetland islands of Bressay and Noss. The Shetland Times Ltd, 2019.

Innocent Passage - the Wreck of the Tanker Braer. With Karen Warner. Mainstream, Edinburgh, 1993.

A *Place in the Sun - Shetland and Oil.* Mainstream, Edinburgh, and ISER Books, St Johns, Newfoundland, 1991.

Biography:

Sixty North to Sixty South – the lives of Jessie and Tammie Laurenson. Amazon, 2022.

Bobby the Birdman – An Anthology Celebrating the Life and Work of Bobby Tulloch. Editor and contributor, with Mike McDonnell. Birlinn, Edinburgh, 2019.

Memoirs, journalism and photography:

To Beijing after Nixon – a grand tour of China in 1972. Amazon, 2023.

That Spring in Prague – a photographic memoir from 1968 (English/Czech edition). Amazon, 2023.

Reporter on the Rocks – memoirs of a recovering journalist. Amazon, 2021.

Old Rock – Shetland in pictures. Shetland Times Ltd, 1990.

Books for children:

The Travels of Magnus Pole. Amazon, 2022 (third edition, first published in 1974 by Canongate).

Magnus Pole Goes West. Amazon, 2022.

Granny Linda and the Lighthouse. Shetland Heritage Publications, 2014 (revised edition, first published in 1976 by Canongate).

Cheer Up, Grampa Gloomifjord! Illustrated by Martin Emslie. Shetland Times Ltd, 2011.

Wilma Widdershins and the Muckle Tree. Illustrated by Gurli Feilberg. Shetland Times Ltd, 2008.